Ecosystems Research Journal

Arctic Research Journal

Ellen Rodger

Author: Ellen Rodger

Editors: Sonya Newland, Kathy Middleton

Design: Rocket Design (East Anglia) Ltd

Cover design: Margaret Amy Salter

Proofreader: Angela Kaelberer

Production coordinator and prepress technician: Margaret Amy Salter

Print coordinator: Margaret Amy Salter

Consultant:

Written and produced for Crabtree Publishing Company by White-Thomson Publishing

Front Cover:

Title Page:

Photo Credits:

Cover: All images from Shutterstock

Interior: Alamy: p. 6b Sean O'Neill, p. 7t Robert Harding, p. 18t Timothy Epp, p. 24l All Canada Photos.
Ron Dixon: pp. 5, 8; iStock: p. 6t QueGar3, p. 7b CoreyFord, p. 13t Mats Lindberg, p. 15b PaulLoewen, pp. 16–17 drferry, p. 16b Murphy_Shewchuk, p. 18b Nnehring, p. 20bl tillsonburg, p. 21b RONSAN4D, p. 23m bsrieth, p. 27m sarkophoto; Shutterstock: p. 4 Avatar_023, p. 5b La Nau de Fotografia, p. 8 Stockdonkey, p. 10b Ionut Catalin Parvu, p. 11t Maksimilian, p. 11b Tobias Brehm, p. 12t FotoRequest, p. 12b Panaiotidi, p. 13b Vladimir Melnik, p. 14b ILYA AKINSHIN, p. 15t ID1974, p. 16t critterbiz, p. 17t Dan Bach Kristensen, p. 19t Martin Hejzlar, p. 19b NaturesMomentsuk, p. 20t captureandcompose, p. 21t Vadim Petrakov, p. 21t khlungcenter, p. 22t Double Brow Imagery, p. 22b Seita, p. 23t Ana Gram, p. 23b Sergey Uryadnikov, p. 24r orxy, p. 25t Madlen, p. 25b Karel Gallas, p.25m Morphart Creation, p. 26t Michelle Holihan, p. 26b mj - tim photography, p. 27t Morphart Creation, p.27b keerati, p. 28 Armin Rose, p. 29 DonLand; Wikimedia: 5tr NASA, p. 9t Lindsay Nicole Terry, p. 9b Doc Searls, p. 10t Paul Gierszewski, p. 14t Timkal.

Library and Archives Canada Cataloguing in Publication

CIP available at the Library and Archives Canada

Library of Congress Cataloging-in-Publication Data

Names: Rodger, Ellen, author.
Title: Arctic research journal / Ellen Rodger.
Description: New York, New York : Crabtree Publishing Company, 2018. |
Series: Ecosystems research journal | Includes index.
Identifiers: LCCN 2017029303 (print) | LCCN 2017030425 (ebook) ISBN 9781427119285 (Electronic HTML) | ISBN 9780778734680 (reinforced library binding : alkaline paper) ISBN 9780778734932 (paperback : alkaline paper)
Subjects: LCSH: Arctic regions--Environmental conditions--Research--Juvenile literature. | Biotic communities--Research--Arctic regions--Juvenile literature. | Ecology--Research--Arctic regions--Juvenile literature. | Arctic regions--Description and travel--Juvenile literature.
Classification: LCC GE160.A68 (ebook) | LCC GE160.A68 R63 2018 (print) | DDC 577.0911/3--dc23
LC record available at https://lccn.loc.gov/2017029303

Crabtree Publishing Company

www.crabtreebooks.com 1-800-387-7650

Printed in Canada/082017/EF20170629

 In Canada: We acknowledge the financial support of the Government of Canada through the Canada Book Fund for our publishing activities.

Published in Canada
Crabtree Publishing
616 Welland Ave.
St. Catharines, Ontario
L2M 5V6

Published in the United States
Crabtree Publishing
PMB 59051
350 Fifth Avenue, 59th Floor
New York, New York 10118

Published in the United Kingdom
Crabtree Publishing
Maritime House
Basin Road North, Hove
BN41 1WR

Published in Australia
Crabtree Publishing
3 Charles Street
Coburg North
VIC, 3058

Contents

Mission to the Arctic

On a map of the Arctic, the polar sea looks solid like land. But it is not. It is moving blocks of ice, not a solid sheet. Those blocks are growing thinner and weaker each year.

My research mission is to study **climate** and **environmental change** in the Arctic. I have been sent by the Arctic Climate Research Institute (ACRI) to see how these changes affect the ocean and the land. They also have a big impact on the animals and people that live in the Arctic. On my journey, I will follow parts of the **Northwest Passage**. This is a path for ships to cross the Arctic. Thick ice once made the journey impossible, even in summer. Many ships were lost trying to find a passage through the ice. Today, melting and thinning ice has opened the passage, bringing more ships to the Arctic each year.

The Arctic region includes the Arctic Ocean and surrounding waters. It is also ringed by land belonging to eight countries. These are the USA, Canada, Russia, Finland, Iceland, Greenland, Norway, and Sweden.

This satellite photograph shows how the Northwest Passage is almost completely ice-free.

Field Journal: Day 1

Iqaluit, Nunavut Territory

I flew to the Canadian territory of Nunavut. My plane landed in Iqaluit, on Frobisher Bay, Baffin Island. The early summer sun shines almost all day and night here. This is called the "midnight sun." It happens because of the tilt of Earth's **axis** as it orbits the Sun. This made it possible to watch the beluga and bowhead whales in the bay even at 10 p.m.

Bowhead whales can live up to 200 years in Arctic waters.

The Arctic is warming more quickly than anywhere else on Earth. Most scientists agree that this warming is because people are using **fossil fuels** such as coal, oil, and gas.

Iqaluit is the largest community in Nunavut, with around 6,700 people. 60 percent of the population are an **indigenous** people called **Inuit**. They hunt and fish for food. I went to a meeting where Inuit elders and youth discussed how to tackle climate change. They said it was making it difficult to find and hunt caribou. Rising temperatures are affecting the animals' food source, so caribou numbers are decreasing. Developing a plan won't stop climate change, but it will help the Inuit adapt to the changes.

Iqualuit means "place of many fish."

natstat STATUS REPORT ST456/part B

Name: Narwhal
(Monodon monoceros)

Description:
Narwhals look like whales with a unicorn horn. The male has a long, spear-like tusk on its head. The Inuit call them Qilalugaq Qirniqtaq, which means "the one that is good at curving itself to the sky." They are an important food source for Inuit peoples. Narwhals live under the sea ice in the Baffin Bay area. Recently, narwhals have started moving farther north, following the ice and the Arctic cod that they eat. These fish need ice to survive.

Attach photograph here

Threats:
Loss of habitat and pollution due to increased shipping in Arctic waters and melting ice.

Numbers:
80,000 in this region.

Status:
Near-threatened.

Field Journal: Day 3

Pangnirtung, Nunavut Territory

The wind blew steadily as I walked through this small Baffin Island community. My guide told me that a few years ago there were heavy rains and a lot of **meltwater**. These wore away the permafrost holding up two bridges. The bridges were damaged and had to be closed. Temperatures that year were the warmest in over 12 years.

The Arctic Circle is an imaginary circle around the top of Earth. It marks the areas where the Sun does not set for six months during summer. The Sun also does not rise for six months during winter.

Permafrost is Arctic soil and rock that is permanently frozen, often for thousands of years.

Arctic sea-ice cover reaches record low

September 1984

September 2012

Pangnirtung is just a few miles away from the Arctic Circle. Almost all the buildings here are built on permafrost. It keeps their foundations solid. But even this far north the warming climate is melting the permafrost. Buildings must be propped up by adding rock, gravel, and steel beams. Without this support they will shift, tilt, and fall apart.

Temperatures in Pangnirtung hit a record of 50°F (10°C) in April 2016. The normal high is 19.4°F (–7°C).

natstat STATUS REPORT ST456/part B

Name: Penny ice cap

Description:
This ice cap is in Auyuittuq National Park. An ice cap is like a small ice sheet. Auyuittuq means "the land that never melts" in the Inuktituk language. Temperatures in the summer have risen higher and higher since 2005. This means that the ice caps and **glaciers** here are melting faster than ever before. The meltwater is causing flooding and rising sea levels around the world.

Threats:
Global warming.

Numbers:
2,200 square miles (5,700 square kilometers).

Attach photograph here

Field Journal: Day 8

Pond Inlet, Baffin Island and Bylot Island

My guide picked me up at Pond Inlet on Baffin Island. He explained that we were going to a nearby "bird island" where there is a bird **sanctuary**. The ice had just broken up, allowing us to make the 16-mile (25-kilometer) boat ride to Bylot Island's Sirmilik National Park. The bird sanctuary here is an important nesting area for birds that migrate, including snow geese and thick-billed murres.

Sirmilik means "the place of glaciers" in Inuktitut. The park has 16 glaciers.

Thick-billed and common murres in their Arctic breeding ground.

There were murres on the shore as far as the eye could see. Their black heads and backs shone in the sun. These birds are expert ocean divers. They can dive as deep as 330 feet (100 meters) in search of Arctic cod. They thrive in cold climates but have had to adapt to the warming ocean. Higher temperatures mean that the ice breaks up earlier than it once did. This means that the murres have to lay their eggs earlier.

Bylot's birds

71 species of birds

35 breeding species

6 year-round species

100,000 greater snow geese nest here

30,000 murres breed here

50,000 kittiwakes nest here

Field Journal: Day 11

Grise Fiord, Ellesmere Island, Nunavut Territory

I have arrived at the top of the world! This is the northernmost settlement in Canada. I saw walruses diving in the **fiord** while I was out walking. My guide from the Hunters and Trappers Organization pointed out some grazing muskoxen. These large animals live in herds. They eat lichens, mosses, and grasses. The Inuit hunt muskoxen for food.

Sightings

I saw a colony of Arctic terns and their fuzzy hatchlings. Incredible! This is their breeding ground and they make the long flight here from Antarctica each year. That's a whopping 25,000 miles (40,000 kilometers)!

Muskoxen and polar bears are also hunted for sport by **big game hunters**. Local people earn a living working at lodges and as hunters' guides. All hunters need permits to kill, and only a few are issued each year. The government tries to control hunting so that not too many animals are killed.

Hunters travel from all over the world and pay to be taken on hunts.

natstat STATUS REPORT

ST456/part B

Name: Polar bear
(Ursus maritimus)

Description:
Polar bears were once threatened by overhunting. Many countries signed agreements to limit the hunting. This has helped the bears survive. Polar bears can swim great distances, but they need sea ice to hunt seals. Seals live on the ice and are the bears' main food source. The polar bears' hunting habitat is shrinking because the sea ice is melting faster.

Attach photograph here

Threats:
Global warming and loss of sea-ice habitat, increased use of bear habitat.

Numbers:
About 26,000 worldwide (no definite figures).

Status:
Vulnerable.

Field Journal: Day 13

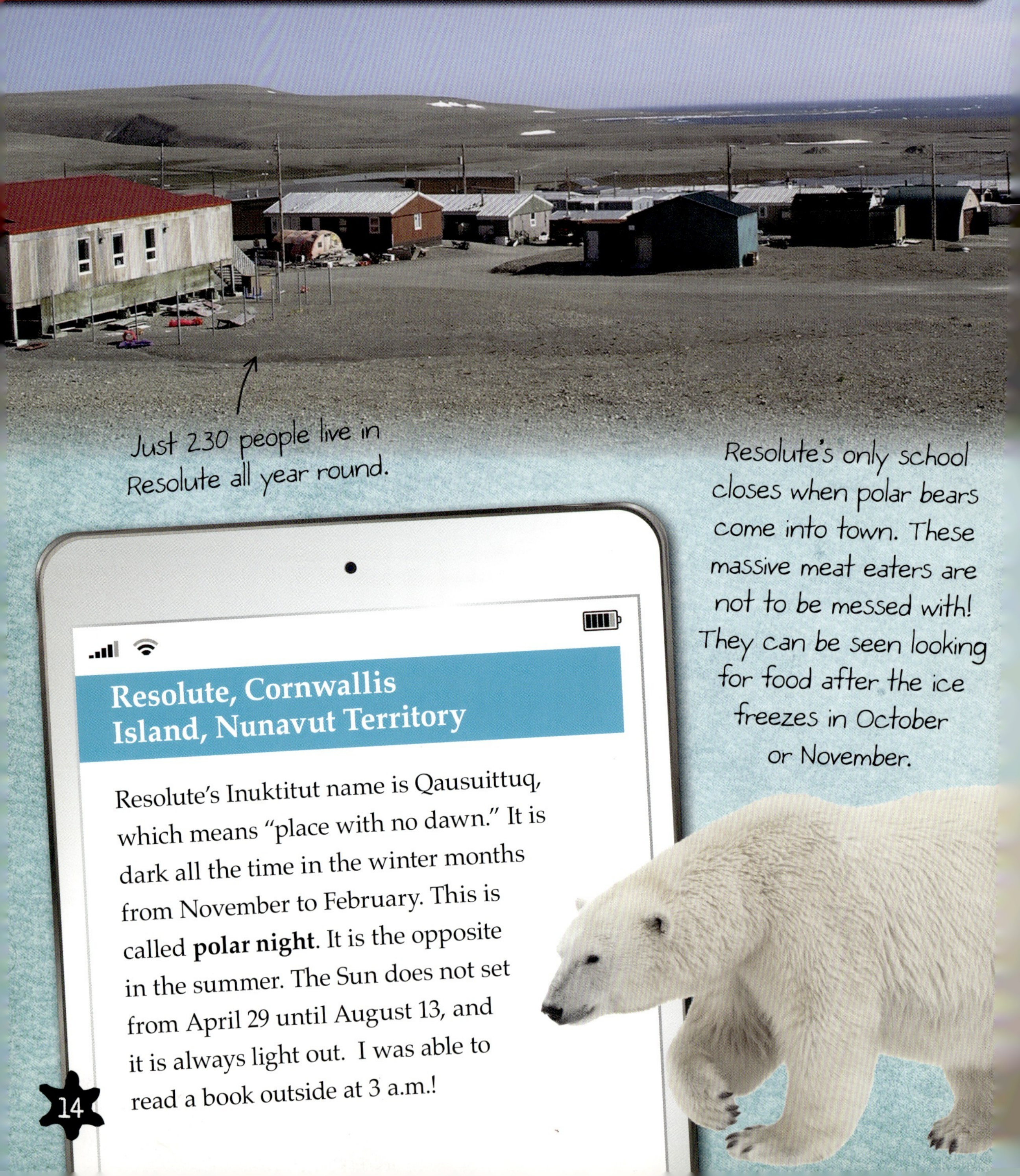

Just 230 people live in Resolute all year round.

Resolute's only school closes when polar bears come into town. These massive meat eaters are not to be messed with! They can be seen looking for food after the ice freezes in October or November.

Resolute, Cornwallis Island, Nunavut Territory

Resolute's Inuktitut name is Qausuittuq, which means "place with no dawn." It is dark all the time in the winter months from November to February. This is called **polar night**. It is the opposite in the summer. The Sun does not set from April 29 until August 13, and it is always light out. I was able to read a book outside at 3 a.m.!

I visited the High Arctic Weather Station at Resolute. This station has been gathering climate information since 1947. Resolute is one of the coldest places on Earth. It has an average yearly temperature of 3.7°F (–15.7°C). It was a brisk 40°F (4.4°C) the day I visited. It felt cold to me, but it was much warmer than usual for this area. Resolute is now a stop on the Northwest Passage route. Melting sea ice means more ships come here.

Weather stations across the Arctic are tracking effects of climate change on this region.

natstat STATUS REPORT ST456/part B

Name: Peary caribou (*Rangifer tarandus pearyi*)

Description:
Peary **caribou** are found only in Canada. This Arctic animal travels from island to island in winter and spring. The number of Peary caribou has dropped over the last 50 years. They eat lichen, sedges, and grasses. In very cold years it can be difficult for them to find food under the snow and ice. In very warm years, melting ice makes it hard for them to move between islands to find food.

Attach photograph here

Threats:
Over-hunting, starvation due to climate change.

Numbers:
10,000–15,000.

Status:
Endangered.

Field Journal: Day 15

Cambridge Bay, Victoria Island, Nunavut Territory

Wildflowers grow knee-high on the tundra near Cambridge Bay. The tundra is a large, treeless plain. My guide on this hike brought a gun for protection in case we met any grizzly bears. They can be dangerous to humans who venture into their territory. But the only thing we shot were pictures! Away in the distance, about 25 muskoxen were munching on lichen.

My guide pointed out an Arctic wolf in the distance.

Explorer Roald Amundsen from Norway was the first to cross the Northwest Passage, in 1905. A ship that once belonged to Amundsen became stuck in the ice at Cambridge Bay and sank. It was recently raised to the surface.

The next day we began a three-day canoe trip on the Ekalluk River. A fish called giant Arctic char enter the river from the Arctic Ocean each August. Recently, fishers have noticed the char flesh is pink rather than red as it normally is. Scientists believe this is because the char are eating more capelin and less pink shrimp. Capelin are a small ocean fish. Capelin adapt and grow better than pink shrimp as the Arctic waters get warmer.

Arctic char are related to salmon. They are a major food source for the Inuit.

Cambridge Bay is the largest stop for ships on the Northwest Passage route. The passage has been ice-free in the summer since 2015. More ships, including cruise ships, are planning to sail through.

Northwest Passage crossings

Explorers have tried to sail the Northwest Passage for hundreds of years. The ice was always too thick and dangerous, and many died. Now the warming sea ice makes the passage possible.

Year	Ships passing through
1909	1 ship
1969	1 ship
1980s	4 per year
2005	7 ships
2010	19 ships
2012	30 ships
2014	17 ships

Field Journal: Day 19

Tuktoyaktuk, Northwest Territory

Next on my itinerary was a day trip to Pingo National Landmark. This lies outside Tuktoyaktuk, or "Tuk" as people who live here call it. Pingos are amazing hills of ice-filled soil. They rise 16–229 feet (5–70 meters) from the permafrost. They were formed thousands of years ago, when frozen ground was forced up by the pressure of water underneath.

Sightings

I spotted some bright white tundra swans feeding. These birds nest on the tundra in the summer and gather on nearby lakes.

Swans

I hiked to the top of Ibyuk pingo. At 164 feet (50 meters), this is the tallest pingo in Canada. It is still growing! There are 1,400 pingos near here. Over long periods of time, the frozen ground under the pingos can melt. Then the cone-shaped mounds collapse into craters. The warming Arctic climate makes them melt and cave in more quickly.

Lemmings are prey for Arctic foxes.

natstat STATUS REPORT ST456/part B

Name: Arctic fox (*Vulpes lagopus*)

Threats: Global warming.

Description: The Arctic fox's pure white winter coat makes it a favorite with fur traders. They are not endangered, even though they are still trapped. The major threat to Arctic fox populations is the red fox. Red foxes could not survive in the Arctic before climate change. Now, they compete for food such as lemmings, which are like mice.

Status: Least concern.

Attach photograph here

Field Journal: Day 21

Inuvik, Northwest Territory

I headed out on an **ATV** to visit some climate scientists who were working at a lake outside town. They were collecting samples of ancient trees found on the bottom of the lake. The trees have been preserved in the very cold water. Their rings tell us about temperatures in the past and provide a record of climate change. Each ring represents a year of the tree's life. Wide tree rings mean a long growing season when there was plenty of moisture.

Caterpillar tracks on the ATV help us get around easily in the snow.

My Inuvialuit guides gave us muktuk to eat. This is frozen whale skin and blubber.

The Inuvialuit people hunt bowhead whales for food. In recent years, there have been more bowheads in the Arctic Ocean. This may be because there is less sea ice in the summer. Whale hunting is good here for the Inuvialuit, but a warmer climate makes it difficult to hunt the seals they eat. Things are changing, so hunters cannot trust their traditional knowledge of the ice any more. They are more likely to get caught in dangerous storms. They may also fall through the ice and drown.

Mosquitoes hatch earlier and grow larger as temperatures increase. They also live longer. Mosquitoes can bother caribou so much that they spend more time running away from the insects than eating. This makes the caribou weak. They fall prey to disease and predators.

natstat STATUS REPORT ST456/part B

Name: Barren-ground caribou (*Rangifer tarandus groenlandicus*)

Threats:
Humans, bears, wolves, climate change.

Description:
Caribou are big, plant-eating mammals. There are eight large herds in the Arctic. They move from the tundra to the **taiga**, traveling up to 750 miles (1,200 kilometers) each year. There are fewer caribou than there once were. Scientists think this is because the plants they eat in the spring grow earlier now. The caribou do not return from their migration early enough.

Numbers:
1.2 million.

Attach photograph here

Field Journal: Day 24

Ivvavik National Park, Yukon Territory

"Look, up there," my guide told me as a gyrfalcon flew above us. These birds are a common sight in Ivvavik National Park. We were hoping to see part of the Porcupine caribou herd in the area where they give birth to their young. Porcupine caribou have the longest migration route of any land animal on the planet.

Sightings

In Ivvavik National Park I saw some dall, or thinhorn sheep. These are horned sheep that roam a nearby mountain range called the British Mountains.

Dall sheep

To protect the Porcupine caribou, not many people are allowed near them. I saw hundreds of the caribou on the coastal plain near the Beaufort Sea as we flew in. Ivvavik is also home to grizzly bears, polar bears, black bears, wolves, muskoxen, and moose.

Moose.

Porcupine caribou are named after the Porcupine River.

natstat STATUS REPORT ST456/part B

Name: Grizzly bear (*Ursus arctos*)

Description:
Grizzly bears are important to the health of **ecosystems** in this part of the world. They are considered a **keystone species**. Grizzlies are known to act aggressively when they feel threatened. Mothers protecting cubs can be especially dangerous.

Attach photograph here

Threats:
Climate change, habitat loss, humans.

Numbers:
6,000–7,000 in the territory.

Status:
Least concern.

Field Journal: Day 27

The Flats are a **wetland** area along the Old Crow River. They contain around 2,700 shallow lakes, ponds, and marshes.

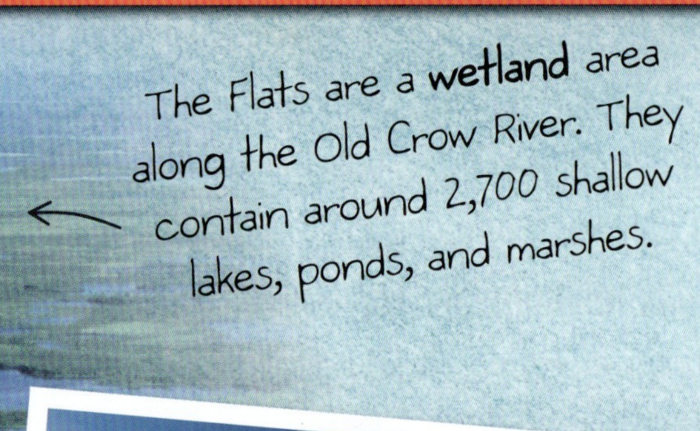
Lupines

Old Crow Flats, Yukon Territory

Plants called Arctic lupines carpet the Old Crow Flats area. The Gwichin people have been hunting and fishing here for thousands of years. They talked to us about the changes they have seen over the last few decades. Temperatures have risen, permafrost is melting, and animal habitats have changed.

I walked with an elder through patches of plants called Labrador tea, cotton grass, and cloudberry. He pointed out where he lays muskrat traps in the winter. He also showed me where the moose and Porcupine caribou herds come to eat. We took a boat out on a lake and saw peregrine falcons, osprey, and two bald eagles. The elder told me that a lot of lakes are drying up. The water in the rivers is also moving more slowly. Some rivers are even changing direction.

Cloudberry

Sightings

I saw teeth and bones in the riverbed that had turned into fossils. Old Crow is the richest region in Canada to find fossils from the last ice age. Extinct animals such as woolly mammoths roamed here 20,000 years ago.

Woolly mammoth

Muskrats are trapped for their meat. Their skin is sold to make fur clothing.

Field Journal: Day 30

Utqiagvik is the most northern city in the USA. The cemetery here is sinking because the permafrost is melting.

Arctic Coast: Utqiagvik (Barrow), Prudhoe Bay, Alaska

I traveled on a boat that dropped off tools and food at Utqiagvik and Prudhoe Bay. Prudhoe Bay is the largest oil field in North America. The flat tundra landscape is dotted with drill sites of oil companies that pump oil from the ground to sell. Thousands of people work for the oil companies in this area. This land is also used by Arctic and red foxes, caribou, grizzlies, polar bears, and migrating birds.

The oil is pumped through pipes to southern Alaska. The Trans-Alaska Pipeline System (TAPS) is one of the world's largest pipelines. Many accidental oil spills have endangered plants and wildlife. The largest oil spill on the tundra occurred here in 2006. A tiny hole in the pipeline let out 267,000 gallons (1,010,705 liters) of oil. The government made efforts to clean it up, but no clean-up will remove all of the spilled oil.

Sightings

We saw some bowhead whales off the coast. These whales migrate around the Arctic region throughout the year, rather than seeking out warmer waters.

Bowhead whale

An oil pipeline cuts through the Arctic landscape.

In 1989, a ship called the *Exxon Valdez* ran aground in the Gulf of Alaska. It spilled millions of gallons of oil.

The spill killed 250,000 seabirds, thousands of otters, and 22 whales called orcas.

Nearly 30 years later, only seven of 28 animal species studied have recovered from the effects of the spill.

Final Report

Some experts believe that there will not be any summer ice at all in the Arctic Ocean within 20 years.

Report to: ARCTIC CLIMATE RESEARCH INSTITUTE (ACRI)

OBSERVATIONS

This trip has been my first visit to so many Arctic communities and environments. It is a huge geographic area with many amazing plants and animals. The Arctic Ocean alone covers more than 5.4 million square miles (14 million square kilometers). That's bigger than all of Europe! Frozen for most of the year, it is rapidly melting. Arctic temperatures are rising at twice the global average.

FUTURE CONCERNS

An ice-free Arctic might be good for international business, but it is not so great for the environment. Many commercial and research ships will be going through the Northwest Passage, disturbing the habitat. The recent discovery of the sunken ships *Terror* and *Erebus*, lost in 1845, will also bring curious tourists. These cruise ships will add to shipping traffic.

Conservation Projects

Global warming remains the biggest threat to the Arctic ecosystem. The indigenous Inupiaq people are helping scientists study changes in the Arctic sea ice. The Inupiaq can supply knowledge of how melting ice and increased ship traffic is affecting the whales. Some plants, animals, and people can adapt to climate change in the Arctic. But they can do do little to stop it on their own. Everyone needs to try to help slow climate change. We can save energy and cut down on our use of fossil fuels. I will have video conferences instead of flying for meetings to remote locations. Governments can put limits on shipping traffic and drilling for oil. We can invest in clean energy to reduce pollution.

To help achieve this, 195 countries signed the 2015 Paris Agreement on Climate Change. These countries agreed to lower the amount of greenhouse gases they emitted and to spend more money on clean energy sources. The goal of these actions is to stop global temperatures from increasing further and to slow the melting of Arctic sea ice. But there remains a lot to do.

The ships that bring tourists to the Arctic cause fossil-fuel emissions that contribute to more melting.

Your Turn

* Use the information in this book to give three examples of how Arctic peoples depend on and interact with their environment. Why do they care about the health of the ecosystems they live in?
* Describe how climate change is altering the Arctic environment and ecosystem.
* How are Arctic animals adapting to a changing climate? Why is adaptation important?

Learning More

BOOKS

In Arctic Waters by Laura Crawford, illustrated by Ben Hodson (Sylvan Dell Publishing, 2013)

Ice Bear by Nicola Davies, illustrated by Gary Blythe (Candlewick Press, 2003)

Polar Oceans by Bobbie Kalman (Crabtree Publishing, 2003)

WEBSITES

www.ngkids.co.uk/places/ten-facts-about-the-arctic
This National Geographic page gathers 10 detailed and informative facts about the Arctic.

http://polardiscovery.whoi.edu/arctic/ecosystem.html
Check out the interactive Arctic Ocean ecosystem illustration from the Woods Hole Oceanographic Institution and learn about the plants and animals that call this ecosystem home.

http://kids.nceas.ucsb.edu/biomes/tundra.html
This useful site is a detailed exploration of aquatic and terrestrial biomes, including the Arctic tundra.

https://climate.nasa.gov/vital-signs/arctic-sea-ice/
This website from NASA contains information about Arctic sea ice, including a fascinating time-lapse sequence showing changes to the ice cover between 1978 and 2014.

Glossary & Index

ATV all-terrain vehicle; vehicles that are designed to travel on surfaces like snow

axis an imaginary line from the top to the bottom of the Earth, around which it rotates

big game hunters hunters who kill animals such as polar bears for sport instead of food

caribou large reindeer that roam northern and polar regions of North America

climate change a change in global climate patterns, such as warmer winters and summers, that is due to increased greenhouse gases such as carbon dioxide in the Earth's atmosphere

ecosystem all the living things in an area that interact with each other

environmental change change that occurs in a specific environment or area

fiord a long, narrow, and deep inlet of sea water that is between two cliffs

fossil fuels natural fuel such as coal or gas that was formed millions of years ago from the remains of living things

glacier a slowly moving mass of ice that was formed millions of years ago near the Earth's poles or near mountain ranges

indigenous original inhabitants of a region, or an animal or plant that is native to an area

Inuit an indigenous people of northern and Arctic Canada, and parts of Greenland and Alaska

keystone species a species that keeps the populations of other species under control

meltwater water that has melted from snow and ice or from a glacier

Northwest Passage a sea passage along the northern, or Arctic, coast of North America that links the Atlantic Ocean to the Pacific Ocean

polar night the constant night experienced in the Arctic polar regions for six months of the year, caused by the tilt of the Earth's axis

sanctuary a safe place for living things, where they are protected

taiga in North America, the barren areas of the boreal forest between the tree line and the tundra

wetland land made up of marshes and swamps